essentials

essentials liefern aktuelles Wissen in konzentrierter Form. Die Essenz dessen, worauf es als „State-of-the-Art" in der gegenwärtigen Fachdiskussion oder in der Praxis ankommt. *essentials* informieren schnell, unkompliziert und verständlich

- als Einführung in ein aktuelles Thema aus Ihrem Fachgebiet
- als Einstieg in ein für Sie noch unbekanntes Themenfeld
- als Einblick, um zum Thema mitreden zu können

Die Bücher in elektronischer und gedruckter Form bringen das Fachwissen von Springerautor*innen kompakt zur Darstellung. Sie sind besonders für die Nutzung als eBook auf Tablet-PCs, eBook-Readern und Smartphones geeignet. *essentials* sind Wissensbausteine aus den Wirtschafts-, Sozial- und Geisteswissenschaften, aus Technik und Naturwissenschaften sowie aus Medizin, Psychologie und Gesundheitsberufen. Von renommierten Autor*innen aller Springer-Verlagsmarken.

Beatrice Marie Ellerhoff

Calculating with quanta

Quantum computer for the curious

 Springer

Beatrice Marie Ellerhoff
Department of Physics and Department
of Geosciences, Tübingen Universtiy
Tübingen, Baden-Württemberg, Germany

ISSN 2197-6708   ISSN 2197-6716   (electronic)
essentials
ISBN 978-3-658-36750-3   ISBN 978-3-658-36751-0   (eBook)
https://doi.org/10.1007/978-3-658-36751-0

Preface

The phenomena of quantum mechanics fascinate not only the research world but increasingly also the general public. The bizarre properties of the tiny building blocks of our matter, called quanta, arouse widespread curiosity: Harnessing quantum properties for computation has the potential to provide quantum computers with computing power exponentially faster than the best high-performance computers of our time. As a result, their development is of great importance to research, industry, and society. This book explores the quantum phenomena of entanglement and superposition and explains how they can be used for calculations. It aims to provide a clear, lively, and descriptive understanding of the processes involved in quantum computing. Starting with the basic encoding of information, through the explanation of simple algorithms, to possible applications, I look forward to guiding curious readers through the world of quantum computing. This little book is dedicated to my father, who sparked a curiosity for science in me. As a young researcher, I am grateful to be given the chance to pass on this enthusiasm, and I thank all those who contributed with their support. My special thanks go to Tim Ellerhoff for his loving support, and to Abra Ganz, Amelie Scupin, Bruno Faigle-Cedzich, Christopher Lance, Shirin Ermis, and Uli Schünemann for their thorough and critical reading of the text.

Heidelberg, Germany   Beatrice Marie Ellerhoff
January 2022

V

What You Can Find in This *essential*

- How quantum computers work
- An illustrative explanation of quantum entanglement
- How quantum particles can be used for computing
- Derivations and simple explanations of quantum algorithms
- The approaches to and significance of quantum error correction
- Current challenges and potential applications of quantum computers

Contents

Photo: Ute von Figura

Beatrice Marie Ellerhoff is a PhD student at the Institute of Environmental Physics in Heidelberg and at the Geo- and Environmental Research Center (GUZ) in Tübingen, Germany. During her studies in physics at Heidelberg University, she wrote for the blog www.manybodyphysics.com and the *Science Notes* magazine. She worked for 4 years as a student assistant at the Max Planck Institute for Medical Research, including in the group of Nobel Prize winner Stefan Hell. In her master's thesis, she investigated the entanglement of many ultracold quantum particles. As part of her doctorate, Beatrice Ellerhoff is now turning her attention to the macroscopically tangible world and researching the fundamentals of climate fluctuations on long time scales in Kira Rehfeld's group. She is particularly fascinated by the linkage of physical concepts across disciplines. In addition to research, she is passionate about running and jazz piano. She was a fellow of the Heidelberg Graduate School for Fundamental Physics and receives a doctoral scholarship from the Heinrich-Böll-Stiftung.

Beatrice Marie Ellerhoff, Geo- and Environmental Research Center (GUZ), Schnarrenbergstr. 94-96, 72076 Tübingen, Germany.

From a physics perspective, quantum mechanics has shaped the past century. In addition to Albert Einstein's general theory of relativity, it was the quantum theory that has led to a paradigm shift in the way we look at the laws of nature. Ever since Niels Bohr presented the first atomic model with quantum mechanical properties in 1913, physicists have been describing the world of the smallest particles using the language of quantum mechanics. In this context, energy and momentum are not continuous, but restricted to discrete "quantized" values. Quantum mechanics goes even further, as it describes not only radiation but also matter as waves.

The modern quantum mechanics created the basis for a multitude of new achievements. It was an astonishingly short path from the discovery of the fundamental principles to their first applications, such as the laser. Now another "quantum leap" is in sight: Quanta will be used to process information and to enable unprecedented computing power. Quantum computers could outperform today's supercomputers by far and thus answer unsolved questions, for example in the field of drug development. Numerous research groups and companies are working intensively to bring quantum computing out of its infancy and promise new insights for many areas of industry and research.

Quantum Revolution

2

2.1   Quantum Mechanics in Our Everyday Life

We encounter classical physics every day. Apples fall from trees. The pressure in a water bottle rises in the glow of the sun. A magnetic compass needle points to the north. These properties belong to macroscopic objects: the apples, the liquid, the needle. They represent an ensemble consisting of numerous atoms and molecules whose microscopic details are not necessary to describe the properties of the object. These simplifications are often very practical in the so-called classical physics. Just imagine that in order to calculate the pressure in a gas cylinder, the behavior of each individual molecule had to be studied. Nevertheless it is worthwhile looking into the world of the smallest building blocks of nature, as we will see in this book.

In contrast to classical physics, quantum physics aims to describe the world of atoms, electrons, or photons. For these tiny quantum particles, the laws of nature are fundamentally different to in classical physics. It is hard to imagine the properties of quanta, but in many respects, they can be compared to waves. Quantum particles can superimpose, amplify or attenuate each other. Moreover, quanta are capable of combining mutually exclusive properties. For example, mathematically speaking, a quantum particle can be in two places at the same time, since its location is only linked to a probability and is therefore not fixed.[1]

---

[1] The double-slit experiment, in which a single electron penetrates two slits simultaneously due to its wave properties, is an impressive demonstration of these properties [8].

© The Author(s), under exclusive license to Springer Fachmedien
Wiesbaden GmbH, part of Springer Nature 2022
B. M. Ellerhoff, *Calculating with quanta*, essentials,
https://doi.org/10.1007/978-3-658-36751-0_2

Despite their complexity, we use quantum mechanical properties every day. The most important examples are lasers. They find universal application in industry and medicine. Moreover, X-ray diagnostics and magnetic resonance imaging are based on quantum effects, as is the modern semiconductor technology found in every smartphone. Last but not least, light-emitting diodes (LEDs) are based on quantum physics. This series of groundbreaking quantum achievements of the last century could be long continued. For this reason, experts speak of the first quantum revolution, which is characterized by the fact that quantum effects are used to enable new technologies.[2]

Researchers are currently working on the realization of a second quantum revolution [3]. This seeks to explore quanta as information carriers and to design a new type of computer, the quantum computer. Individual states as well as the European Union are investing massively in its development. Numerous research groups and large IT corporations have joined the race.

Surprisingly, the initial idea was not to use quanta as computational units, thereby building a computer. Rather, the question was: "How can the complicated world of quanta be better understood with the help of computers?" More than 40 years ago, Nobel Prize-winning physicist Richard Feynman claimed in his lecture "Simulating physics with computers" that calculations of the quantum world are too complicated for ordinary computers. As we are overwhelmed by the complexity of quantum mechanics, so are today's computing centers (Fig. 2.1). Richard Feynman had a revolutionary idea: couldn't quanta themselves be used to build new computers? In other words, is it possible to build computers that exploit the laws of quantum mechanics to solve quantum problems?

Indeed, Richard Feynman was right, as first prototypes of quantum computers have been developed in the past two decades (Fig. 2.2). They promise the solution of problems for which today's supercomputers are too slow. These include understanding chemical reactions and the structure of complex molecules, analyzing large amounts of data, describing phenomena in weather research, and encoding communication. This will advance the application of quantum mechanics in our everyday life.

---

[2] Nature has long been ahead of mankind in using quantum effects. For example, the chemical process of photosynthesis, in which light energy is converted into chemically bound energy, is based on quantum effects.

Fig. 2.1 Nowadays, classical supercomputers solve difficult tasks in industry and research. The picture below shows the design of the "Karlsruhe High Performance Computer" (HoreKa for short), launched in mid-2021. It is among the 15 most powerful computers in Europe and has the computing power of more than 150,000 laptops. But even for these "super machines", some tasks remain difficult. This is where future quantum computers come into play. (Source: KIT Steinbuch Centre for Computing (SCC), Information Technology Centre (ITC) at the Karlsruhe Institute of Technology (KIT), www.scc.kit.edu)

Fig. 2.2 Quantum computers do not look like computers, but resemble a variety of all kinds of physical instruments, including lasers and vacuum pumps. These are needed to control the quantum particles and thus to be able to use them for calculations. (Source: IQOQI Innsbruck/M. R. Knabl)

2.2   Towards Unimaginable Computing Power

But what exactly makes quantum computing so fascinating? Why does a single quantum computer have the potential to solve tasks that entire computing centers fail at today?

Calculating with quanta is a bit like conducting a big band. The conductor leads the piece, but each instrument, from the trumpet to the saxophone, follows its own line. The sound is created by the interplay of the instruments and the superimposition of all notes. The melody of a single instrument might sound illogical or even boring, and hardly reveals the main piece. Sometimes an improvisation occurs. Here the rhythm section, consisting of bass, drums, guitar, and piano, spontaneously accompanies a solo instrument. At this very moment, the music is created solely by the way the musicians interact with each other. It is not created by the sound of a single instrument alone, but by an ensemble of harmonies, rhythms, and melodies. The true musical information is hidden in the overlap and interplay of the instruments.

The same applies to information processing in quantum networks. Quantum computers can perform several computational steps simultaneously, just as the melody of the trumpet can run parallel to that of the saxophone. Similar to this auditory impression, created by the sound of many tones, the quantum information is contained in the superposition of quantum states. Because of the constant mutual influence of quantum bits (qubits), quantum computers can compute with all qubits at the same time. The following chapter discusses how this can be explained from the perspective of quantum mechanics and how quantum algorithms can be programmed.

There is, however, a catch: neither quantum computers nor the improvising musicians can tell what they are actually doing at that very moment. If the soloist is disturbed during improvisation, errors creep in, other musicians are thrown out of sync and the whole performance might be ruined. Errors propagate furiously in quantum networks. Small disturbance can stop the calculation immediately. This makes quantum computers very fragile and difficult to control, which is currently the biggest challenge. Many research groups focus solely on how to avoid or correct errors.

But what are the "instruments" of quantum computing? What elements is the "quantum big band" made of? Unlike conventional computers, which are almost exclusively composed of electronic semiconductor devices, the future quantum hardware remains uncertain. Some quantum computers are based on electrically charged atoms (ions) near the absolute freezing point (-273.15 °C). Moreover, they may be based on superconductors that can conduct electricity without any loss. In addition, there are a variety of approaches using other quantum particles such as electrons, photons, or anyons. In Chap. 4 we give an overview of promising approaches for the technical implementation of quantum computers. It remains to be seen which of these techniques will prevail, and until then it remains unclear how the quantum music of the future may sound like.

The Basic Building Blocks of Quantum Computing

3

3.1   From Bits to Quantum Bits

Information in the digital world, from pocket calculators to aircraft control software, has one thing in common: it can be represented by sequences of zeros and ones. The language of today's computers is based on so-called bits ("binary digits"), representing the numbers "zero" and "one". These two possible values of a bit are mutually exclusive, like "yes" or "no", "north" or "south", or like a light switch that can be either "on" or "off", but never both at the same time. One bit thus represents the elementary unit of information and its change the smallest possible calculation step in conventional data processing.

However, incrementally storing and passing zeros and ones can require vast computing resources. For example, storing the word "hello" overwrites 35 bits of a normal computer. The more complex the task, the greater the amount of memory and computing power required. The evaluation of large data sets in artificial intelligence is therefore already being outsourced to supercomputing centers. It requires typically a lot of energy to run these centers. Nevertheless, they might take days, weeks or even months for a particular calculation. Some problems, such as decoding molecular structures, cannot be solved exactly even by these computers.

Some experts expect that the trend to produce smaller computer chips while at the same time increasing their performance is limited [13, 14]. Today, transistors used in computer chips are barely larger than 10 nanometers – a fraction of a body cell. The physical laws at these scale are different from classical mechanics. Once the size of molecules and atoms is reached, quantum effects start playing an

© The Author(s), under exclusive license to Springer Fachmedien
Wiesbaden GmbH, part of Springer Nature 2022
B. M. Ellerhoff, *Calculating with quanta*, essentials,
https://doi.org/10.1007/978-3-658-36751-0_3

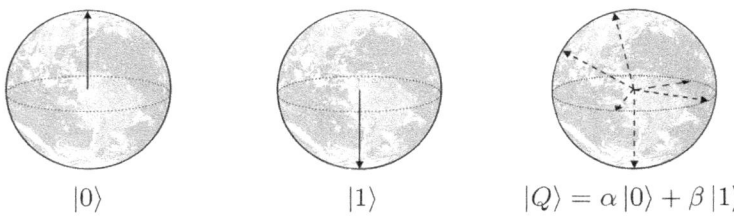

Fig. 3.1 A quantum bit $|Q\rangle$ can assume any superposition between "zero" $|0\rangle$ and "one" $|1\rangle$. Figuratively speaking, it can point not only to the "north" or "south" but also to any point on Earth's surface. (Source: Illustration by the author)

important role. These effects hinder the design of conventional computer chips because they can interfere with the electronic circuits. Quantum computers could remedy this problem in the future, as they promise to surpass today's computing capacities by far.

Quantum computers are also based on the states "zero" and "one", similar to their classical counterparts. However, the states are encoded in particles with quantum mechanical properties, such as atoms, electrons, or photons (see Sect. 4.1). That's why they are called quantum bits (qubits). According to the laws of quantum mechanics, a qubit state can not only be "zero" or "one". In addition, these states can overlap. The qubit value is then in a mixture of "zero" and "one". Thus, it can contain all (infinitely many) states in between "zero" and "one".

One can illustrate this superposition by means of the Earth's surface (Fig. 3.1). The quantum states "zero" and "one" correspond to the opposite poles "north" and "south" [12]. While a classical bit represents only the states "north" or "south", a quantum bit (marked with $|Q\rangle$) can point to any point on the globe by assuming a superposition of "north" and "south".

The bracket notation $|\cdot\rangle$, which is typical for quantum mechanics, is called Dirac notation.[1] The state of a qubit, that is the superposition of "north" $|0\rangle$ and "south" $|1\rangle$, can then be expressed as:

$$|Q\rangle = \alpha\,|0\rangle + \beta\,|1\rangle$$

---

[1] This notation goes back to the physics Nobel Prize laureate Paul Dirac (1902–1984).

α and β are (complex) numbers that determine the proportion of $|1\rangle$ and $|0\rangle$. They provide information on the length and angle of the arrow, indicating the qubit state on the sphere. It is important to note, that this state is located on the surface of the sphere, thus satisfying the normalization $|\alpha|^2 + |\beta|^2 = 1$ according to Pythagoras' theorem. Precisely, $|\alpha|^2$ and $|\beta|^2$ show the probability of finding the particular states

$|0\rangle$ and $|1\rangle$, respectively. One possible qubit state is $|+\rangle = \dfrac{1}{\sqrt{2}}|0\rangle + \dfrac{1}{\sqrt{2}}|1\rangle$, for

example. It consists of equal shares of $|0\rangle$ and $|1\rangle$. On the globe, this corresponds to an arrow that points in the direction of the equator (cf. Fig. 3.2). Analogous to

the state $|+\rangle$, there is also a superposition $|-\rangle = \dfrac{1}{\sqrt{2}}|0\rangle - \dfrac{1}{\sqrt{2}}|1\rangle$. It differs only by

the minus sign and points in the opposite direction along the equator.

To determine a qubit's state, measurements are required. The measurement process can be imagined by looking at the qubit through a slit (Fig. 3.3) [7]. Inspecting the $|+\rangle$–state in this way, one finds "north" in half the cases and "south" in the other half, as $|\alpha|^2 = 50\%$ and $|\beta|^2 = 50\%$ indicate the probabilities of finding the qubit value "north" or "south". The slit makes the superposition state $|+\rangle$ "collaps" into one of the two possible values.

While classical computers use binary digits, quantum computers rely on probabilities, related to quantum mechanical superpositions. One calculation step brings the qubit from one superposition to another. As long as the qubit value is not measured, it remains undetermined. Several computational paths can run at once. This procedure is already much more efficient than the exchange of information between conventional bits.

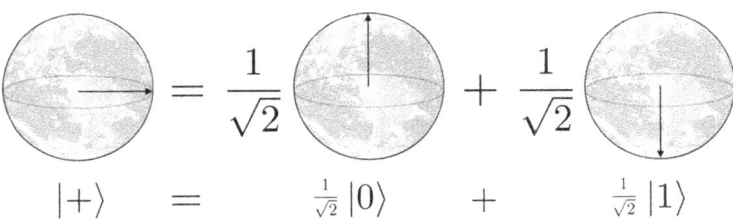

Fig. 3.2 A perfect superposition of $|0\rangle$ and $|1\rangle$ corresponds to the quantum state $|+\rangle$ and points, figuratively speaking, towards the equator of the globe. (Source: Illustration by the author)

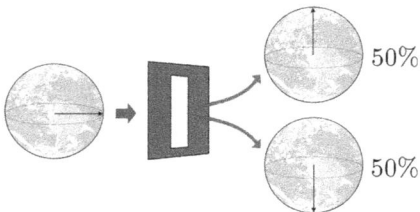

Fig. 3.3 Measuring the "equator" state |+⟩ along the north-south axis results in "north" or "south" with equal probability. (Source: Illustration by the author)

But there is yet another quantum mechanical property that tremendously increases the potential of quantum computers: the phenomenon of entanglement. Entanglement enables the members of a qubit network to influence each other as if by internal coupling. This makes the quantum computer, at its core, a state of overlapping and mutually dependent qubits. As we will see in the following chapter, such a quantum computer no longer simply passes information from bit to bit. Instead, it computes with all qubits at the same time. Theoretically, quantum computers are therefore equiped with a decisive advantage over today's computers.

3.2   Quantum Advantage Through Entanglement

Entanglement is considered to be the most characteristic phenomenon of the quantum world. Researchers[2] such as Erwin Schrödinger, Albert Einstein, Werner Heisenberg and Niels Bohr debated this phenomenon and its physical interpretation early on. This is because entanglement seems to violate previously valid laws of nature. It enables quantum particles, for example electrons, to be connected to each other over arbitrarily long distances as if by an invisible band, and to influence each other instantaneously. Albert Einstein called this "spooky" action at a distance. Approximately 80 years later, researchers were able to entangle photons from a station on Earth to a satellite 1200 km away [20]. In such an experiment, the particle on Earth "knows" the properties of the distant counterpart. If someone on Earth wants to reveal these properties, it is sufficient to measure only the particle on Earth due to the entanglement.

---

[2] Between the men listed, there was a lively debate about the foundations of quantum mechanics. But were there only famouse male researchers? Marie Curie should be mentioned as outstanding physicist, however, she was not involved in the mentioned debate. Lucy Mensing was a pioneer in quantum mechanics whose career ended too soon. A definitive answer requires an investigation of the discrimination of women in science [15].

As an example, consider an entangled pair of qubits

$$|Q_A, Q_B\rangle = \frac{1}{\sqrt{2}}\left(|0,0\rangle + |1,1\rangle\right). \tag{3.1}$$

This is a superposition of two states in which both particles have the same value. In the first state, both qubits have the value "zero" $|0, 0\rangle$. In the second state, they have the value "one" $|1, 1\rangle$. This is a so-called Bell pair, featuring the highest degree of entanglement. The knowledge about a single qubit is sufficient to know the state of the other one [2].

Experimentally, a Bell pair can be measured as follows: If physicists want to determine the value of the Bell pair $|Q_A, Q_B\rangle$, the pair must decide between the superposition of $|1, 1\rangle$ and $|0, 0\rangle$. This decision is enforced by the measurement of a single qubit and can be compared to the observation through a slit (Fig. 3.4). Even if only qubit A *is* forced into a state by the measurement, B instantaneously assumes the same value, even if B is infinitely far away. This is the essence of entanglement: the information is contained in all entangled qubits and the change of one immediately affects all others. If qubit A *is* forced into a particular state, qubit B instantaneously assumes the same value, even if B is infinitely far away. This is the essence of entanglement: the information is contained in all entangled qubits and the change of one qubit immediately affects all others.

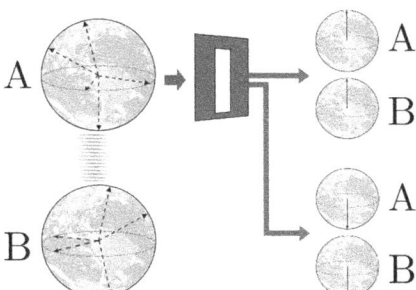

Fig. 3.4 Measuring a qubit A entangled with B reveals not only the value of A, but also the value of B. Even if the qubit B is far away and does not see the slit, the measurement instantaneously changes the value of its entangled counterpart. In this example, the Bell state $|Q_A, Q_B\rangle = \frac{1}{\sqrt{2}}\left(|0,0\rangle + |1,1\rangle\right)$ is decomposed into its constituents $|0, 0\rangle$ and $|1, 1\rangle$ with equal probability. (Source: Illustration by the author)

This fast and constant exchange of information greatly increases the computing power of quantum computers for many mathematical problems. The information transfer from bit to bit, which is necessary for a computation step in classical computers, is not needed anymore due to entanglement. All qubits perform the computation step simultaneously. The network thus carries more information than the sum of its parts. The more elements the network comprises, the more states can be processed simultaneously and the more computing power is available. If a pair of qubits is used instead of a single qubit, the performance doubles. With ten qubits, the computing power is already increased by more than a factor of 1000. This is an exponential increase. Quantum computers with only a few dozen qubits thus have an unimaginably higher efficiency compared to conventional computers (see Fig. 3.5). A quantum computer with about 50 qubits can, in theory, outperform the computing power of today's supercomputing centers. This enormous computing potential is only due to the bizarre properties of quanta. However, the very same properties also make it challenging to understand the complex processes in quantum computers.

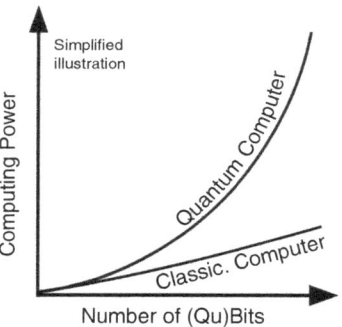

Fig. 3.5 The computing power of quantum computers exceeds that of conventional computers, at least in theory. Unlike conventional computers, it increases exponentially rather than linearly with the number of bits. Since 2019, when Google demonstrated that a quantum computer could solve a very specific task faster than any supercomputer, experts have been arguing about whether the quantum advantage has been physically confirmed [1, 16]. It thus remains unclear whether, and, if so, for what kind of tasks, quantum computers will show an exponential advantage in computing power. (Source: Illustration by the author)

3.3   Quantum Gates and Algorithms

After describing the smallest computational units (qubits) and their superposition and entanglement, the following section focuses on the information processing. In conventional computers, computational problems, such as searching a database, are solved by algorithms that combine the individual computational steps. In a computer chip, these computational steps are processed by so-called logic gates that change the states of individual bits. Gates are comparable to calculation rules such as "plus (+)" or "equal (=)". Analogous to a handwritten calculation, which is built up of initial numbers, arithmetic signs and a result, a digital calculation (algorithm) consists of input bits, gates and output bits. These algorithms run permanently in the background on every laptop. The gates, i.e. the calculation rules, are implemented by components such as transistors.

In the same way, a quantum algorithm computes with qubits and quantum gates on quantum computers. However, quantum gates, unlike classical logic gates, represent a time-controllable interaction of qubits with each other or with the environment, for example, through a measurement. After becoming familiar with how gates work, we can put them together to perform quantum computations. In the following, we will give examples of a few simple quantum algorithms.

Example 1: Entanglement of Two Qubits
The Hadamard and the controlled-NOT gate, also called CNOT, are the most important quantum gates. They repeatedly entangle qubits with each other – a process that is permanently required for efficient quantum computing. A simple quantum algorithm is the creation of entanglement between two qubits, for example, a Bell pair $|Q_A, Q_B\rangle = \frac{1}{\sqrt{2}}(|0,0\rangle + |1,1\rangle)$. The Hadamard gate changes the state of a qubit from a pure "zero" or "one" to a mixture consisting equally of "zero" and "one". Figuratively speaking, the arrow on the globe moves from the pole to the equator (see Fig. 3.6).

Mathematically, these changes can be expressed as follows

$$|0\rangle \xrightarrow{H} \frac{1}{\sqrt{2}}(|0\rangle + |1\rangle) = |+\rangle \quad \text{and} \quad |1\rangle \xrightarrow{H} \frac{1}{\sqrt{2}}(|0\rangle - |1\rangle) = |-\rangle.$$

Since these equations can easily become confusing in case of long quantum algorithms, a graphical representation is used in which qubits are connected to each

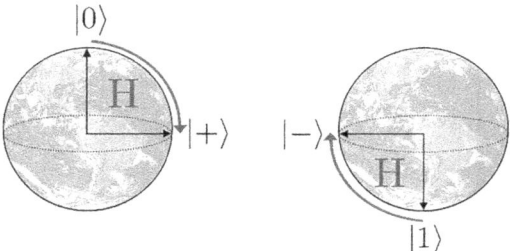

Fig. 3.6 The Hadamard gate changes the pure qubit state "zero" or "one" into a superposition. On the globe this corresponds to a rotation of 90° towards the equator. (Source: Illustration by the author)

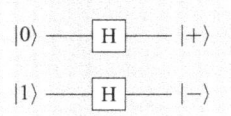

Fig. 3.7 The Hadamard gate transforms the pure state of a qubit into a superposition. In the graphical notation, the states |0⟩ and |1⟩, and |+⟩ and |−⟩ are connected by "cables" onto which gates can be attached. (Source: Illustration by the author)

other as if by cables. The gates are then represented by building blocks that can be inserted between the qubits (cf. Fig. 3.7).

The second building block for creating the entanglement of two qubits is the CNOT gate. In contrast to the Hadamard gate, it imposes conditions on the qubit pair: If qubit A has the value |0⟩, nothing is changed. If qubit A has the value |1⟩, then qubit B is flipped (Fig. 3.8). Flipping means that the qubit takes the opposite value: "zero" becomes "one" and vice versa.

The combination of both building blocks creates the entanglement of two qubits, more precisely, it produces the Bell state $(|Q_A, Q_B⟩ = \frac{1}{\sqrt{2}}(|0,0⟩ + |1,1⟩)$ (see Fig. 3.9). Initially, both qubits were in the "zero" state, $|Q_A, Q_B⟩ = |0, 0⟩$. The Hadamard gate first brings one of the qubits into superposition |+⟩. The CNOT gate measures this superposition state, which is "zero" or "one" with equal probability. Depending on how "the dice fall", the state of the second qubit is changed or not.

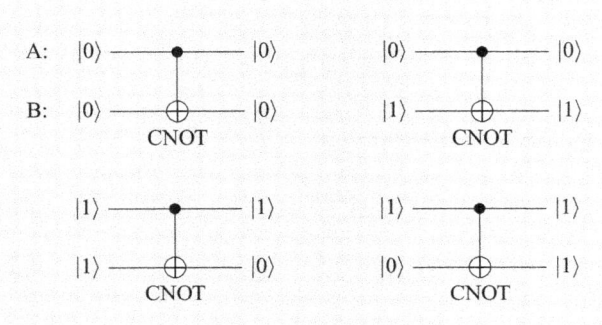

Fig. 3.8 The CNOT gate performs a change of qubit B if qubit A has the value $|1\rangle$. (Source: Illustration by the author)

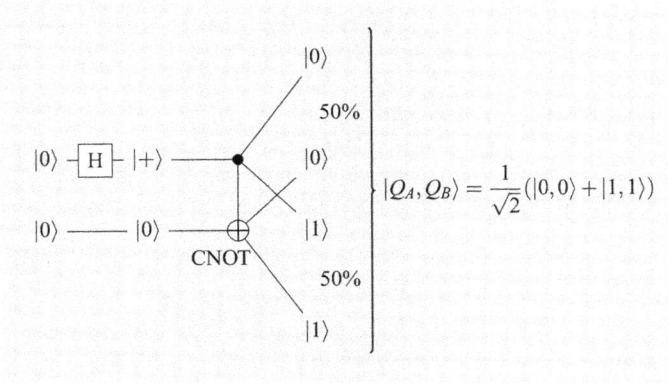

Fig. 3.9 The entangled Bell state $|Q_A, Q_B\rangle = \frac{1}{\sqrt{2}}(|0,0\rangle + |1,1\rangle)$ is generated by applying a Hadamard gate and a CNOT gate to the state $|Q_A, Q_B\rangle = |0, 0\rangle$. (Source: Illustration by the author)

This finally results with equal probability in the superposition of the qubit pairs $|Q_A, Q_B\rangle = |0, 0\rangle$ and $|Q_A, Q_B\rangle = |1, 1\rangle$. As a result, the final state is the Bell state we already know $|Q_A, Q_B\rangle = \frac{1}{\sqrt{2}}\left(|0,0\rangle + |1,1\rangle\right)$.

Example 2: Quantum Teleportation
Another application is the teleportation of quantum information, which is important for quantum encryption and the quantum internet. Teleportation sounds extremely complicated, but requires hardly more gates and computational steps than the entanglement of two qubits. However, here eleportation does not mean the transport of a qubit from one place to another, but the transfer of information from one qubit to another. On a normal computer, this application is comparable to copying a file from one folder to another. However, according to the so-called no-cloning theorem [19], it is not possible to perfectly copy the quantum state of a qubit to another one without changing the initial qubit. However, this is not necessarily a disadvantage. Rather, any attempts to copy qubits leave traces. Unintentional interception of information can thus be traced or even prevented. Therefore, there are great hopes that quantum teleportation will one day enable more robust encryption techniques. Section 4.3 will take a closer look at this.

Quantum teleportation can be realized as follows: Imagine a friend possessing the secret qubit G with the unknown state $|Q_G\rangle = \alpha|0\rangle + \beta|1\rangle$. This state is to be teleported. Conveniently, we possess a qubit that can receive the state and another one that can help with the transmission. Both are transferred to the Bell state $|Q_A, Q_B\rangle = \frac{1}{\sqrt{2}}\left(|0,0\rangle + |1,1\rangle\right)$ using the algorithm for entangling two qubits described above (cf. in Fig. 3.10). We now send the auxiliary qubit A to our friend, where it interacts with the foreign qubit through a CNOT gate. As a result, our remaining qubit B is now also entangled with G. After a Hadamard transformation, our friend measures both qubits. We receive a call with the measurement result. Since our qubit B was entangled with A and G, we know that the measurement changed our qubit state. With some math, we figure out that we just need to rotate our qubit state just a little bit to copy the secret qubit's state. This rotation is done using an X and Z gate, which simply swap $|+\rangle$ and $|-\rangle$, and $|0\rangle$ and $|1\rangle$, respectively (Fig. 3.11). By following the instructions in Table 3.1, we manage to transfer the state of the secret qubit G to our qubit B. The information has thus been copied with the help of the qubits and has been teleported to us with the help of quantum entanglement.

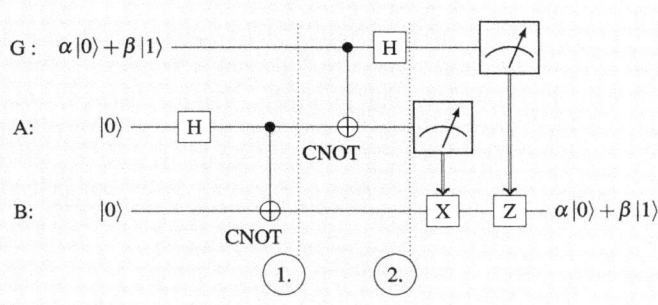

Fig. 3.10 Quantum teleportation protocol to transfer the qubit state $|Q_G\rangle = \alpha|0\rangle + \beta|1\rangle$ to qubit B. An entangled Bell pair is first generated and then entangled with G. After the measurement, indicated by the pointer symbol, the X and/or Z gate is used, depending on the measurement result, to teleport the state (cf. Table 3.1). As a result, qubit B assumes the initial state of the secret qubit G. The state can now be measured at qubit B and thus decoded. (Source: Illustration by the author)

Fig. 3.11 The X gate reverses the pure states $|1\rangle$ and $|0\rangle$ of a qubit. The Z gate, on the other hand, changes the superposition state from $|+\rangle$ to $|-\rangle$, or vice versa. (Source: Illustration by the author)

Table 3.1 Last step of a protocol for teleportation of a qubit state. After a measurement of the qubits A and G, specific quantum gates are applied to qubit B

Measurement result		Gate
A	G	B
0	0	None
0	1	Z
1	0	X
1	1	X Z

The measurement result, communicated by our friend, was the key to receiving the secret information of the qubit G. Since this type of teleportation enables versatile possibilities for the transmission, storage, and processing of qubits, it is an important building block for any application of quantum computers. In particular, novel encryption methods and the quantum internet rely on quantum teleportation. Moreover, it is relevant for the correction of errors in qubit networks, which is currently still the biggest challenge in building powerful quantum computers and will be discussed in the following chapter.

Further Reading

Further explanations, well-researched articled and podcasts related to quantum computing can be found at www-quantamagazine.org.

The Wikipedia articles on "Bell's inequality", "Bell's state" and "Bloch sphere" are recommended to get a deeper understanding of superposition and entanglement.

Another very helpful reading in this context is the book "Quantum Computing for the Quantum Curios" by Ciaran Hughes et al., published by Springer.

Quantum Computing Today and Tomorrow

4

4.1   The Search for the Optimal Hardware

There are different approaches to implement a qubit, the heart of every quantum computer. They must fulfil the following requirements: First, the method of choice must be able to reset qubits to their initial state, similar to deleting a line in a calculator. Second, they must be able to store and retrieve information. It must also be possible to apply gates to them and read out the state of a qubit in a measurement. Lastly, the qubits must obey the laws of quantum mechanics. They should not only offer two different states ("zero" and "one"), but also be able to enter superpositions of these.

A common approach for the physical realization of qubits are superconducting quantum circuits. The phenomenon of superconductivity relies on materials that have no electrical resistance below a certain temperature and thus conduct electricity without loss. However, they often require extremely low temperatures. Typically, superconductors operate slightly above $-200\,°C$. These temperatures can be achieved in experiments using liquid nitrogen. However, there are also promising candidates for so-called high-temperature superconductors. For example, in 2019, such a conductor was found to operate at $-23\,°C$ at the Max Planck Institute for Chemistry in Mainz [4].

At sufficiently low temperatures, electrons form pairs in the superconducting material. If two parts of the conductors are separated by an insulator, there is typically no electric current. However, if the electrons form pairs, they can "tunnel" through such an insulating layer and generate the electric flow [9]. This so-called Josephson effect is equally likely to happen in both directions of the insulator. Brian D. Josephson was awarded with the Nobel Prize in Physics in 1973 for the

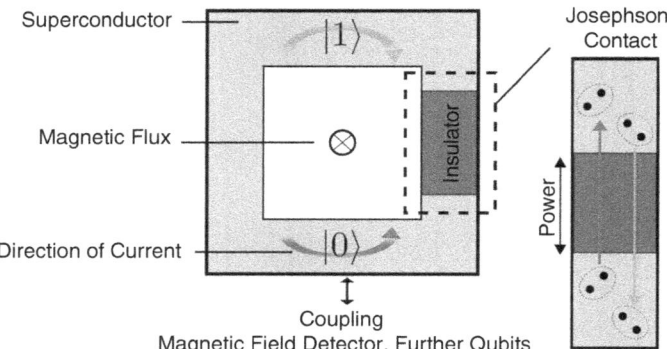

Fig. 4.1 A Josephson junction forms the main part of a superconducting qubit. Electron pairs can "tunnel" through the barrier of the insulator due to the Josephson effect and thus generate an electric current in both directions simultaneously. Magnetic fields and microwave radiation are used to encode the information in the direction of flow. In this way, the qubit state can be controlled and read out. (Source: Illustration by the author)

discovery of this effect. An electrical contact that exploits the Josephson effect is called a Josephson contact. As the electric current can flow in both directions simultaneously, Josephson contacts provide ideal conditions for forming a qubit.

Therefore, a superconducting qubit (Fig. 4.1) typically consists of a superconducting ring with one or more Josephson junctions. Magnetic fields and microwave radiation control the current flow so that the states "zero" and "one" are encoded in a specific direction, for example "$|1\rangle$= clockwise" and "$|0\rangle$= counterclockwise".

However, a single superconducting qubit does not make a quantum computer. It is coupled to other qubits and detectors via electromagnetic fields. These components are placed on a circuit board similar to classical computer hardware. A major advantage of this method is that it makes use of the standard semiconductor technology.

Another technique for quantum computing is the trapped ion quantum computer (Fig 4.2). Ions are atoms or molecules that are electrically charged. Since ions are quantum particles by nature, they can directly act as qubits and their characteristics can be used for calculations. To this end, it is necessary to control the ions, which is done by trapping them in a vacuum chamber. Electrodes generate electric fields

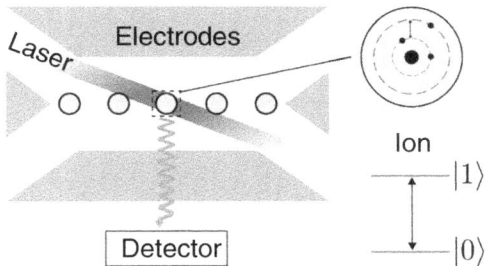

Fig. 4.2 In an trapped ion quantum computer, ions are confined in space by electric fields. Their quantum state can be manipulated by laser pulses. Here, the excited state, in which an electron is further away from the nucleus on average, corresponds to the state |1⟩. During the decay into the ground state, |0⟩, energy is released in form of a photon which is detected by the camera. (Source: Illustration by the author)

that confine the ions in space because of their charge. During this process, they can be cooled down to extremely cold temperatures close to absolute zero, that is $-273.15\,°C$, using laser cooling.[1] Without this "shock freezing", the ions would not be controllable due to thermal motion. After the laser cooling, the ions are lined up in their ground state |0⟩. If the laser then excites one ion, it enters the state |1⟩. In physical terms, one electron is now further away from the atomic nucleus on average. In order to measure the state, the electron is forced to change back to a closer "orbit" by emitting a photon. This fluorescence can be detected with a simple camera and measures the qubit state.

Among the candidates for future quantum computers discussed here, there is a particularly exotic one: the topological quantum computer [10]. It exploits the bizarre properties of the quantum world like no other. Its qubits consist of anyons, which belong to a class of quasiparticles that occur in two dimensions (i.e., they "live" on a surface). Quasiparticles are a group of physical phenomena that behave like particles but are not. They often arise from the collective excitation of a material and the common properties of many quantum particles. For example, they can describe holes in crystal lattices or oscillations in solids. They are often related to the topology, that is the geometric properties, of matter. The electron pairs from the

---

[1] Steven Chu, Claude Cohen-Tannoudji, and William D. Phillips were awarded the Nobel Prize in Physics in 1997 for their cooling of quantum particles using laser light.

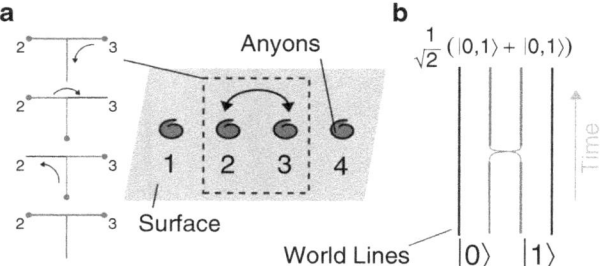

Fig. 4.3 The interchanging of anyons forms braids of their so-called world lines and constitutes a quantum operation. This happens, for example, when transporting anyons along a T-shaped node (**a**). The entanglement of anyons 2 and 3 corresponds to applying the Hadamard gate, which brings the anyon qubit into a superposition state (**b**). (Source: Illustration by the author)

example of superconducting qubits discussed above are also quasiparticles, since they represent a bound state of several particles that behaves like a single particle.

Anyons stand out among quasiparticles due to a special property: Their quantum state reveals something about the course of their recent interactions. This is because their so-called worldlines form braids when anyons pass around one another. Worldlines are paths that document how the particle moves in space and time. Anyons "remember" exactly how these braids were formed. This means that they incidentally provide information about past computating steps. This "memory" can be very handy for detecting errors and preventing their propagation. Better control and correction of naturally occurring errors is considered one of the milestones towards universal quantum computing and will be discussed in detail in the following Sect. 4.2. Another advantage of worldline braiding is that perturbations do not change the state of anyons unless they are strong enough to generate new worldlines, which is rare. But how can such braiding take place?

Imagine a T-shaped node of superconducting nanowires (Fig. 4.3). Anyons are located at the end of these nanowires. Using electric fields, the anyons can be displaced around the wire. This spatial (topological) change can be understood as a quantum gate. The corresponding operation on the world lines encodes the information into the anyons.

Topological quantum computing is still in its infancy compared to the previously mentioned techniques. The theoretical understanding of anyons and the idea of using them for computations is still new. Yet, there is a lack of research expertise to draw on. Nevertheless, topological quantum computing is extremely promising. Since every computational steps is documented in braids, this technique promises easier detection and correction of errors. It could help master the challenge of efficient error correction and decisively advance the development of fault-tolerant quantum computers.

Further Reading
The technology of trapped ion quantum computers has been mainly developed at the Institute of Quantum Optics and Quantum Information (IQOQI) of the Austrian Academy of Sciences [5].

A nice explanation of topological quantum computers can be found on the webpage "https://www.qutube.nl/". In a series of videos they also discuss why a certain class of anyons, so-called Majorana fermions, are often used for topological quantum computers. They are a particularly interesting group of particles, as they represent their own antiparticles.

In their laureate speeches, Nobel laureates William D. Phillips and Brian D. Josephson explain the invention of laser cooling and the discovery of the Josephson effect. They can be found at https://www.nobelprize.org/prizes/physics/.

4.2   The Decoherence Problem

On the one hand, the controlled use of entanglement enables the enormous computing power of quantum computers. The tight interaction of qubits makes them particularly efficient. On the other hand, errors also propagate faster and qubits are not infrequently susceptible to disturbances that change their state. To use a musical analogy: if one instrument in a "quantum big band" plays wrong, it does not only skew up the overall sound, but also disturbs other band members. The harmony of a perfectly assembled orchestra is lost.

Mistakes occuring in quantum calculations can often be describes as a phase or bit flip. While the bit flip reverses the state (e.g. $|0\rangle \rightarrow |1\rangle$), the phase flip changes a sign (e.g. $\alpha|0\rangle + \beta|1\rangle \rightarrow \alpha|0\rangle - \beta|1\rangle$). In the metaphor of the globe, bit flips swap north and south, and phase flips swap west and east. In addition, amplitude errors can occur in which the absolute values of the factors α and β change so that the arrow no longer reaches the surface of the globe, but ends inside the earth. Due to this change of state, the qubit loses its initial information. The former advantage through superposition and entanglement is gone. This problem, called decoherence, is currently the biggest challenge in transforming simple quantum computers to even more powerful machines.

As discussed in the previous chapter, yet, quantum particles can only be controlled under specific conditions, such as extremely cold temperatures. However, this also means that tiny changes in the environment lead to a loss of control. Electrical and magnetic perturbations, insufficient vacuum, temperature fluctuations or vibrations are enough to change qubit states and destroy entanglement – decoherence occurs. Of course, today's research groups are working intensely to construct qubits with states that are as stable as possible. The frequency of errors can be reduced in this way, but so far they cannot be avoided. This is because qubits cannot be completely shielded from the environment. In order to use them for calculations, an information exchange between the qubits and the environment must happen during the measurement process at the latest.

The more qubits are added to a quantum computer, the more likely decoherence occurs. For a long time, quantum computing was therefore considered technically unfeasible. After all, qubits cannot simply be measured to determine whether they are faulty. The physicist Peter Shor found a solution to this problem in 1994 with an ingenious idea [17]. He came up with an algorithm that can be used to effectively repair qubit states. Physicists breathed a sigh of relief and research into correction methods received a boost. Entangle the potentially fault qubit with further auxiliary qubits turned out to be particularly helpful. Because of the entanglement, the auxiliary qubits are able to describe the initial state of the qubit. The states can be compared to detect errors (Fig. 4.4). However, determining the type of error (bit flip, phase flip, or amplitude error) and correcting for it requires further steps. The code in Fig. 4.5 shows how a qubit can be protected against a bit flip with the help of four additional qubits. This example already illustrates the high effort required for many quantum error correction methods: As the number of qu-

Fig. 4.4 Identification of incorrect qubit states: Initially, all three qubits carry the same state. After a potential error occurred, the qubit states are compared. A green light shows that the qubit states are the same, a red light means that they deviate. This pattern reveals where the error is located. (Source: Illustration by the author)

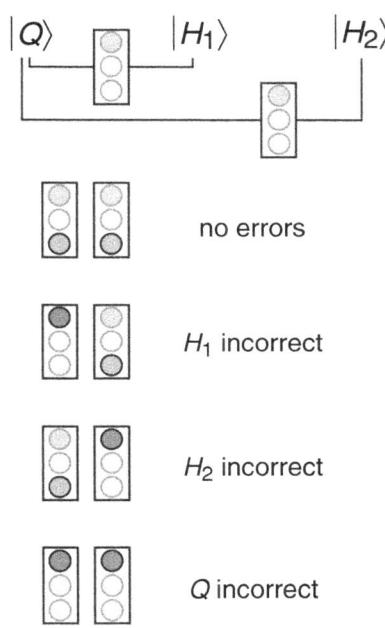

bits increases, the complexity of the computing architecture and thus the error-proneness of the system also increases.

Quantum computers with approximately 50 qubits could theoretically outperform today's best supercomputers. In practice, however, this proves difficult. The Internet giant Google achieved a milestone with its 53-qubit chip "Sycamore" in 2019 [1] (Fig. 4.6). It solved a very specific task faster than today's supercomputers. The question of whether this has already proved the technological superiority of quantum computers has been debated ever since [16]. The distant goal of constructing robust and universal quantum computers that can solve universal tasks will require further effort in terms of qubit stability and software solutions for error correction. Many experts expect that medium-sized systems with about 100 qubits will be realized in the next decade. Novel error correction methods and different quantum computer types, such as the topological quantum computer, promise further progress in this field.

Controlling decoherence and effectively correcting errors poses a challenge to the research community. However, it does not reduce the worldwide efforts to further develop quantum computers. On the contrary, if these problems can be solved, quantum computers could indeed provide an exponential advance in computing power for many computational tasks. What is hoped for from this breakthrough and what the use of quantum computers might look like in the future is discussed in the concluding section.

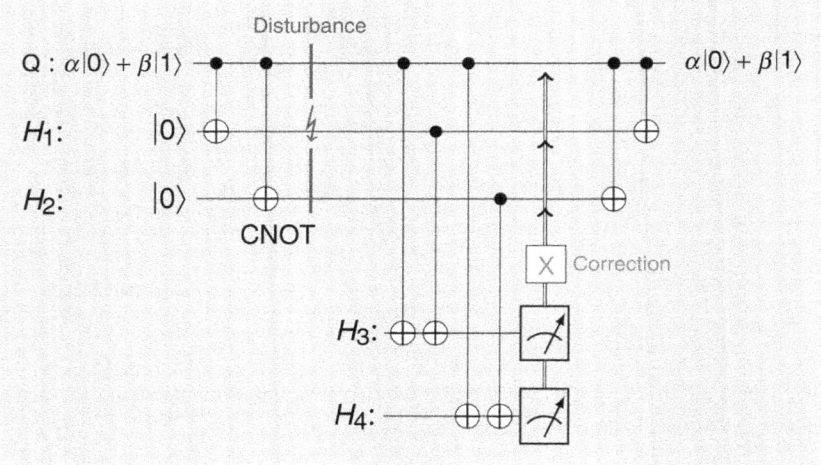

Fig. 4.5 The quantum state $\alpha|0\rangle + \beta|1\rangle$ is passed on robustly with respect to a bit flip by means of four auxiliary qubits. First, the qubit Q is entangled with H_1 and H_2 by two CNOT gates. They are now in the state $\alpha|0, 0, 0\rangle + \beta|1, 1, 1\rangle$. An error occurs and one qubit experiences a bit flip: $\alpha|1\rangle + \beta|0\rangle$. To determine which qubit has changed, four CNOT gates perform the "traffic light" comparison described in Fig. 4.4. The result of the comparison is encoded into two more auxiliary qubits. H_3 indicates whether Q and H_1 carry the same state and H_4 shows whether Q and H_2 are identical. H_4 and H_3 can now be measured without disturbing the system again. The measurement result reveals which qubit is faulty. A correction using an X *gate*, which flips the state again, is applied specifically to the faulty qubit. Finally, the entanglement of the three qubits Q, H_1 and H_2 is removed using two CNOT gates to restore the initial states. (Source: Illustration by the author according to [18])

Fig. 4.6 The "Sycamore" chip is the most important component of Google's latest generation of quantum computers. (Source: Google LLC.)

4.3   The Future of Quantum Computing

Who doesn't think of a standard laptop when they hear the word "computer"? Some may also have the image of a PC with monitor in mind. Or a tablet. Only few might think of huge, cooled halls of refrigerator-like devices– today's super computing centers. This is because laptops, PCs and tablets were designed for everyday use. They can be used to edit documents, cut videos or create presentations.

Highly complex tasks, however, such as the automatic evaluation of thousands of images, bring conventional computers to their limits. Their memory and computing power are insufficient for this. Therefore, these tasks are outsourced to large servers. The internet and cloud computing technology make it possible to have these tasks solved by a data center on the other side of the world. These enormous computing capacities are increasingly in demand as more data is produced and processed in times of digitalization, Big Data, and artificial intelligence.

Yet even the performance of today's supercomputers is limited. For some tasks, such as decoding large molecular structures, they simply take too long. This is where quantum computing comes into play. What questions could quantum computers answer in the near future with their increased computational speed? What needs to be considered in terms of security?

Experiments suggest that one of the first applications of quantum computers will be solving quantum problems themselves. Processes involving quantum particles are similar to the computational processes in quantum computers and can thus be simulated more easily. Fundamental research increasingly no longer relies on simulating quantum problems on classical computers in an elaborate way. Instead, research could simply watch the quantum computer at work – a new research field of quantum simulation has emerged. Processes from particle physics that remain insufficiently understood, such as the generation of matter from light in the so-called Schwinger process, have already been investigated [11]. Similarly, chemical reactions or the properties of molecules could be simulated in the near future. For example, a research group from Innsbruck was able to calculate the energy of molecular hydrogen using a quantum computer [6]. Performing such calculations for more complex molecules or even materials can have many advantages. Research into new drugs and vaccines, as well as more efficient batteries and solar cells, could be facilitated. The discovery of better catalysts that can control chemical reactions under low energy consumption would be attractive not only for industry but also for the environment. In face of the global climate crisis, chemical processes that can effectively remove CO_2 from the atmosphere and bind it in solid carbon (C) are much needed.

Another natural application of quantum computers is the analysis of large amounts of data. Thanks to their ability to test several solutions at once, they are particularly suited to finding an optimal solution from numerous possibilities. Examples are the optimization of traffic control or the decoding of genomes in biological research. Another area of application lies in astronomy, where countless images and signals are evaluated to explore dark matter. With the advent of artificial intelligence and the digitization of many fields, there will certainly be no shortage of large amounts of data that quantum computers could analyze in the future.

Even more complicated than evaluating large amounts of data is making reliable predictions from it. One day, quantum computers might even be able to provide forecasts of price developments on the financial market or play a role in modelling the weather and climate.

Novel encryption and decryption methods and their impact on the security of data streams are often in focus when discussing the abilities of quantum computers. Today, emails, credit card payments or electronic files are mostly encrypted using the so-called public-key cryptography. It is based on the multiplication of long prime numbers. If two prime numbers are chosen, their product can be quickly calculated. However, if only the product is known, it is almost impossible to guess the individual factors. To be precise, today's computers would need decades to do this. No hack is worth this effort!

Quantum computers, however, have the potential to crack this encryption. Estimates suggest that a few thousand to one million qubits are needed to extract the prime numbers in a sufficiently short time. Even though this scenario is presumably far in the future, an IT race for the security of our communication has been sparked. The first alternatives to the public key method have been developed and are considered "quantum safe" for the time being. However, this only means that to date no code tailored for quantum computers is known to crack them.

However, quantum computers can not only decrypt information, but can also be used for encryption. After all, what makes communication safe? On the one hand, attacks should be prevented. On the other hand, people should know whether their information has been attacked or not. Quantum cryptography combines both: Since the message is encoded in quantum states, it is destroyed in the event of an attack. So, if a quantum-encrypted message arrives, it is virtually certain that it was not intercepted. Quantum cryptography is therefore considered to be absolutely tap-proof.

Fig. 4.7 The "IBM Q System One" is one of the most modern quantum computers. IBM makes some of its models available to industry and research. However, these machines can also be accessed by anyone online, in order to carry out initial computing experiments and thus learn how to program on quantum computers. (Source: IBM)

From today's perspective, some possible applications of quantum computers are not foreseeable and are yet to be discovered. The first quantum computer prototypes with up to 20 qubits are available online so that industry and research can try them out (Fig. 4.7).

Certainly, we won't be carrying a quantum computer around like a smartphone in the near future. Nor will we use it like a laptop in our everyday lives. Instead, it has the potential to crack the really "hard nuts" among various research questions. Groundbreaking achievements can be expected in fundamental research, the development of vaccines and drugs, the efficiency of industrial processes, and digital communications. Even though it might take years to construct a universal, fault-tolerant quantum computer, a second revolution in the field of quantum technology is already underway.

Summary

The tremendous potential of quantum computers, which this book attempts to explore and explain, lies hidden in their smallest components, the qubits. They can be physically realized in different ways, for example, using ions, superconductors or anyons. The phenomena of entanglement and superposition allow qubits to interact with each other and share information at an enormous speed. Quantum gates and algorithms control these processes and cleverly use them to solve computational tasks. However, as the number of qubits and computing potential increases, quantum computers become more prone to errors. The development of fault-tolerant hardware and efficient methods for quantum error corrections are currently the greatest challenge on the road to universally applicable quantum computers.

Should future quantum computers be able to surpass the performance of supercomputing centers, they would become preeminently important for the study of complex research questions. Quantum computing has thus the potential to shape many areas of research, industry, economy, and society and to join the multitude of revolutionary achievements in quantum mechanics.

© The Author(s), under exclusive license to Springer Fachmedien
Wiesbaden GmbH, part of Springer Nature 2022
B. M. Ellerhoff, *Calculating with quanta*, essentials,
https://doi.org/10.1007/978-3-658-36751-0_5

- The fundamental differences between quantum and conventional computers
- Explanations of quantum entanglement and superposition
- Quantum algorithms for entanglement and teleportation of quantum states
- An overview of the current state of quantum computing
- Ideas for the implementation of quantum error correction
- Outlook on problems that could be solved by quantum computers in the future

© The Author(s), under exclusive license to Springer Fachmedien
Wiesbaden GmbH, part of Springer Nature 2022
B. M. Ellerhoff, *Calculating with quanta*, essentials,
https://doi.org/10.1007/978-3-658-36751-0

Glossary

Algorithm Sequence of instructions used to solve a specific task, typically given to a computing device.

Anyon A class of quasiparticles that exist in two dimensions.

Decoherence Describes the phenomenon of losing the quantum mechanical superposition of states, for example due to external perturbations.

Electrode Electrical conductor that enables the exchange of electric charges between two media (chemistry) or generates an electric field (physics).

Entanglement Describes the common state of multiple quantum particles that cannot be decomposed into separate states.

Gate Elementary operations to change the value of the underlying unit of information (bits/quantum bits).

Ion Electrically charged atom or molecule.

Josephson effect Tunneling of quantum particles carrying electric charges through a layer between two superconductors.

Quantum bit (Qubit) Smallest possible unit of information in quantum computing.

Quantum Computer Data processing device based on the laws of quantum mechanics.

Quantum Internet Network of connected quantum processors, which can be compared to the internet, but has not yet been realized.

Quantum Teleportation Transmission of information stored in quantum states.

Quasiparticles Excitations in quantum systems that arise from collective properties of multiple particles, but can be described as a single particle.

© The Author(s), under exclusive license to Springer Fachmedien Wiesbaden GmbH, part of Springer Nature 2022
B. M. Ellerhoff, *Calculating with quanta*, essentials,
https://doi.org/10.1007/978-3-658-36751-0

Supercomputer High-performance computers that stand out due to their design, number of processors, and performance.

Superconductor A conductor that transports electricity without electrical resistance.

Superposition Quantum state that arises from the sum of the states of individual quantum particles.

Topology Mathematical study of the properties of geometric objects and spaces.

References

Arute, F. et al: Quantum supremacy using a programmable superconducting processor. Nature **574**, 505–510 (2019). https://doi.org/10.1038/s41586-019-1666-5

Bell, J. S.: On the Einstein Podolsky Rosen paradox. Physics Physique Fizika (1964). https://doi.org/10.1103/PhysicsPhysiqueFizika.1.195

Bell, J.S.: Speakable and Unspeakable in Quantum Mechanics: Collected Papers on Quantum Philosophy. Cambridge University Press (2004)

Drozdov, A.P. et al: Superconductivity at 250 K in lanthanum hydride under high pressures. Nature **569**, 528–531 (2019). https://doi.org/10.1038/s41586-019-1201-8

Georgescu, I.: Trapped ion quantum computing turns 25. Nature Review Physics 2, **278** (2020). https://doi.org/10.1038/s42254-020-0189-1

Hempel, C. et al.: Quantum Chemistry Calculations on a Trapped-Ion Quantum Simulator. Physical Review X **8**, 031022 (2018). https://doi.org/10.1103/PhysRevX.8.031022

Heusler, S., Dür, W.: What one can learn about quantum physics from the single qubit. PhyDid-A **1/11**, 1–16 (2012)

Jönsson, C.: Electron interference at several artificially produced fine slits. Z. Physics **161**, 454–474 (1961). https://doi.org/10.1007/BF01342460

Josephson, B. D.: Possible new effects in superconductive tunnelling. Physics Letters vol. 1 **no. 7**, 51–253 (1962). https://doi.org/10.1016/0031-9163(62)91369-0

Kitaev, A. Y.: Fault-tolerant quantum computation by anyons. Annals of Physics, **303**, 2–30 (1997). arXiv:quant-ph/9707021

Kokail, C. et al: Self-verifying variational quantum simulation of lattice models. Nature **569**, 355–360 (2019). https://doi.org/10.1038/s41586-019-1177-4

Lierta, A.C., Demarie, T., Munro, E.: Quantum computation: a journey on the Bloch sphere. Quantum world association (2018). https://medium.com/@quantum_wa. Cited on 01 Nov 2019

Mack, C.: Fifty Years of Moore's Law. IEEE Transactions on Semiconductor Manufacturing (2011). https://doi.org/10.1109/TSM.2010.2096437

Moore, G. E.: Cramming More Components Onto Integrated Circuits. Proceedings of the
    IEEE (1998). https://doi.org/10.1109/JPROC.1998.658762
Münster, G.: (K)eine klassische Karriere? Physics Journal 19, **No. 6**, (2020)
Pednault, E.et al: Leveraging Secondary Storage to Simulate Deep 54-qubit Sycamore Cir-
    cuits (2019). arXiv:1910.09534v2
Shor, P. W.: Algorithms for quantum computation: discrete logarithms and factoring. Pro-
    ceedings 35th Annual Symposium on Foundations of Computer Science, 124–134
    (1994). https://doi.org/10.1109/SFCS.1994.365700
Steane, A. M.: A tutorial on quantum error correction. Proceedings of the International
    School of Physics "Enrico Fermi". (2006) https://doi.org/10.3254/1-58603-660-2-1
Wootters, W., Zurek, W.: A single quantum cannot be cloned. Nature **299**, 802–803 (1982).
    https://doi.org/10.1038/299802a0
Yin, J. et al.: Satellite-based entanglement distribution over 1200 kilometers. Science **356**,
    1140–1144 (2017). https://doi.org/10.1126/science.aan3211